D1511129

FORCE AND MOVEMENT

Barbara Taylor

Photographs by Peter Millard

FRANKLIN WATTS
New York ● London ● Sydney ● Toronto

Design: Janet Watson

Science consultant: Dr Bryson Gore

Primary science adviser: Lillian Wright

Series editor: Debbie Fox

The author and publisher would like to thank the following children for their participation in the photography of this book: Ozkan Aziz, Terry Cook, Corinne Crockford, Louise Fox, Daniel Kinsey, Thato Matebane, Daniel Seager, Stephen Wilkins and Rosemary Williams. We are also grateful to Julia Edwards and Evelyn Mildiner

Thanks to Heron Educational Ltd for loaning equipment for use in the experiments and Lillywhites Ltd for the loan of sports equipment.

Illustrations: Linda Costello

Franklin Watts Inc.
387 Park Avenue South
New York
NY 10016

Library of Congress Cataloging-in-Publication Data

Taylor, Barbara, 1954-
    Force and movement/Barbara Taylor.
    p.cm—(Science starters)
    Summary: Photographs, simple text, and activities introduce the subject of force and the way things move.
    ISBN 0-531-14081-4
    1. Motion—Juvenile literature. 2. Force and energy—Juvenile literature. [1. Motion. 2. Force and energy.] I. Title.
    II. Series.
QC73.4.T38   1990
531'.112—dc20   89-21505
                                            CIP
                                            AC

Printed in Belgium

CONTENTS

This book is all about the pushing and pulling forces that make things start moving, stop moving, slow down or change shape. It is divided into four sections. Each has a different colored triangle at the corner of the page. Use these triangles to help you find the different sections.

▼ These red triangles at the corner of the tinted panels show you where a step-by-step investigation starts.

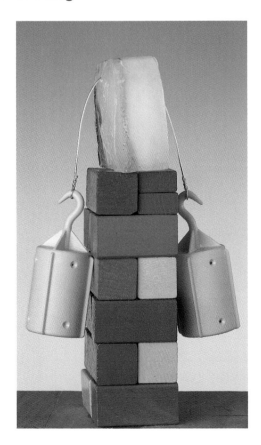

PUSH AND PULL

Have you ever flown a kite?
What makes the kite move?

The wind pushes the kite up into the sky. You pull on the strings to make the kite twist and turn or to bring it down to the ground. Pushes and pulls like these are called forces. We cannot see forces, but we can sometimes see the effect they have on things around us. And we can feel our muscles getting tired when we push or pull things.

Forces can make an object start moving or stop moving. They can also make an object that is already moving speed up, slow down or change direction. It takes more force to move heavy objects. Even though they are trying hard, these boys cannot exert a strong enough pulling force to lift the heavy bag off the ground. Forces do not always make things move.

When two people on roller skates push against each other, they both move backward. This is an important point about forces. Whenever there is a force on something in one direction, an equal force is acting on something else in the opposite direction.

Forces such as squeezing, squashing, twisting, pressing, stretching or bending make things change shape. A car crusher has enough force to squash the empty shell of an old car almost flat. This makes it easier to cut up the metal into fist-sized pieces, which can be used to make new cars.

To make bread, you have to fold and squash the dough. This is known as kneading. It mixes the yeast and flour so the bread rises properly. It also makes it easier to push and pull the dough into different shapes.

STOP AND START

A baseball does not move unless you drop, throw or hit it with a bat. When an object is not moving, it stays still, unless a force acts on it. A moving baseball will not stop until you catch it or it hits the ground. When something is moving, it keeps moving unless a force makes it stop or change direction.

When a heavy demolition ball swings against a building, it produces a force strong enough to knock down the building.

All sports show examples of forces in action. In judo, people need to use a lot of force against their opponents' force to get them moving. But once they are moving, their bodies keep moving until they meet an equal force – the ground.

How much force is needed to start things moving? To measure this, ask a friend to stand on a skateboard or roller skates with a spring balance tied to the waist. Pull on the balance and take a reading when your friend starts to move.

Stand in a line with some of your friends. Hold on to the waist of the person in front and run forward in a straight line. Ask someone else to shout "STOP" to the first person in the line.

Even though you try hard to stop, you will fall over the person in front.

The same thing happens when a car stops suddenly. The passengers are all thrown forward. If they are not wearing seat belts, they may hit the seat in front, or even go through the windshield. Seat belts stop this happening, because they push backward as your body pushes forward.

STICK AND SLIP

When one thing slides over another, a force called friction tries to stop the movement. This is because all surfaces (even those that look smooth) are covered with tiny bumps and holes. When two surfaces touch, these bumps and holes "stick" together.

Friction can be very useful in our everyday lives. Without it, we would slip over every time we tried to walk and we could not fasten things together. It would also be very difficult to pick things up or hold them.

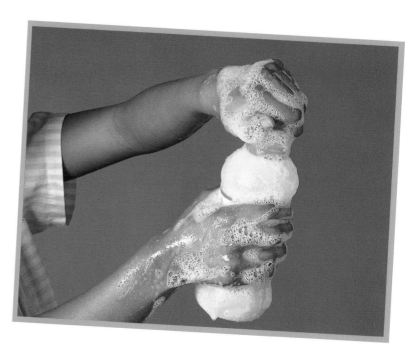

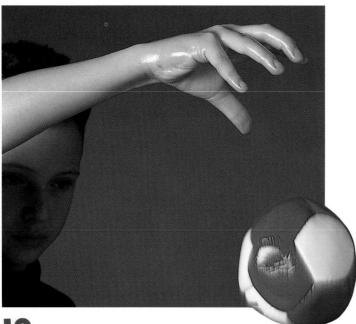

Spread some soap on your hands and try to open a bottle or jar. It's not easy is it!

Can you hold a soccer ball when your hands are covered in cooking oil?

In both cases the friction has been reduced.

To find out more about friction, try this test.

1 Make a slope by propping up a board on some books.
2 Collect some small, flat objects, such as a coin, a plastic button, an eraser, a metal pencil-sharpener and a wooden block.

3 Ask a friend to help you hold the objects in a row at the top of the slope. Let them all go at the same time. Friction between the objects and the board will slow down some objects more than others. Which one reaches the bottom first and which is the last?
4 Change the angle of the slope and repeat the test.
5 Make the slope out of different materials, such as carpet, tiles, rubber or sandpaper. On smooth surfaces, the objects slide more easily than on rough surfaces.

There is more friction on a rough surface than on a smooth one. If you look at the bottom of a bath or shower you may see a bumpy surface. This helps to keep us from slipping.

You can measure the amount of force needed to overcome friction on different surfaces and you can make things move.

1 Push some pins into one end of a wooden board and balance a pencil between the pins. The pencil should be able to turn.

2 Attach a small hook into one end of a wooden block.

3 Tie some yarn or string to the hook and attach a plastic dish to the other end.

4 Place the wooden block at the opposite end of the board to the pencil. How many marbles do you need to put into the dish to make the block move?

5 Now pin some sandpaper to the board and repeat the test. How many marbles do you need to make the block move over the rough sandpaper? What happens if you cover the block in sandpaper too?

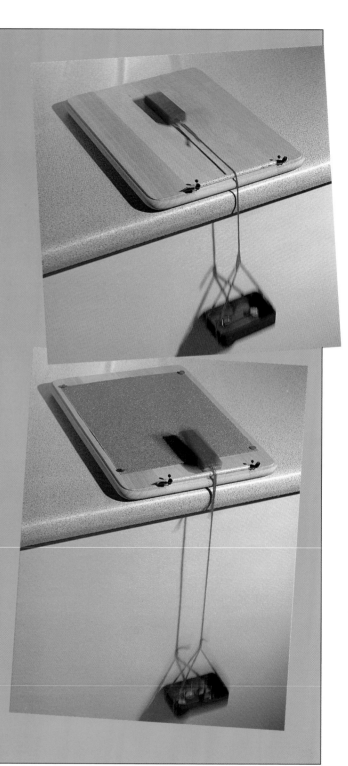

The surface of a road is rough, which increases the friction between vehicle tires and the road surface. This helps the tires to grip. The road surface near a traffic circle or traffic lights is often rougher than the rest of the road. This helps the cars to stop quickly.

Which of these soles would grip best on a smooth, slippery surface?

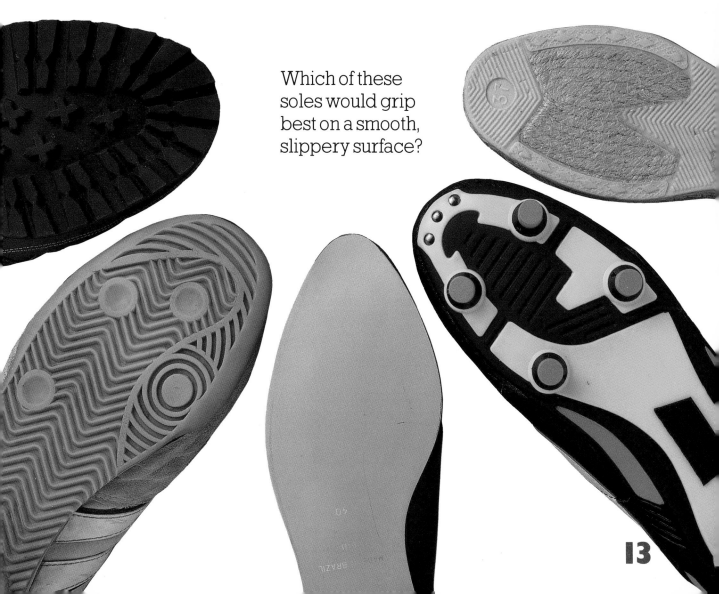

Try sliding a book along a surface. Then put some marbles under the book and slide it again. The marbles stop the book rubbing against the surface, which makes it move more easily.

In many machines, ball bearings do a similar job. They stop surfaces rubbing together and help mechanical parts to turn around easily. This boy is investigating a giant model of ball bearings.

Friction is a big problem inside machines. As the moving parts rub against each other, they become hot and wear out. A layer of oil can be used to keep the machine parts from sticking together. You need to oil a bicycle for the same reason. Oil reduces friction and is called a lubricant.

Have you ever been down a water slide? Water is another lubricant that makes things slide past each other more easily. You can zoom down a water slide much faster than you can down an ordinary slide.

A hovercraft glides along on a cushion of air. This lifts it off the ground or water. The two surfaces are not touching, so there is very little friction to slow down the movement of the hovercraft.

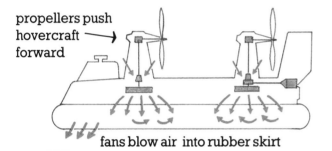

propellers push hovercraft forward

fans blow air into rubber skirt

To see how a hovercraft works, try this test.

1 Find a shallow polystyrene food tray. You can decorate the sides if you like.

2 Use a sharp pencil to make a hole in the middle of the tray.

3 Blow up a balloon and hold the end tightly. Push the end through the hole in the tray.

4 Put your hovercraft on a smooth, flat surface, give it a gentle push and it should begin to glide away. Can you hear and feel the air escaping and pushing the hovercraft along?

PRESSURE

Snowshoes help to keep people from sinking into the snow. They are wide and flat and this spreads the downward push of their feet over a bigger area. The amount of force on a certain area is called pressure. A sharp knife cuts better than a blunt one because there is more pressure on the cutting edge. With a blunt knife, the force is spread over a greater area.

Shoes with thin, pointed heels concentrate all their pushing force into one tiny area. There is a high pressure at that point, which means they sink further into soft sand.

You can use pressure to melt ice.

1 Leave a container of water in the refrigerator until it turns to ice.

2 Tie some heavy objects of equal weight to either end of a piece of thin wire.

3 Balance the ice on a tower of bricks and place the wire over the ice with one weight on each side.

4 The pressure of the wire will melt a narrow channel through the ice. The water in the channel above the wire should freeze back into ice again, because there is no longer any pressure to melt the ice.

Ice skaters glide easily across the ice. The whole weight of the skater pushes down on the narrow blades of the skates, which means there is a lot of pressure on a small area. This force melts the ice and reduces the friction between the metal blade and the ice. The skater slides easily over the ice on a thin layer of water.

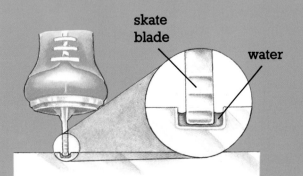

skate blade

water

SIMPLE MACHINES

Machines are not just big, noisy, complicated things like washing machines or cars. Scissors and nutcrackers are machines too. Here are some examples of simple machines. All complicated machines are made up of many simple ones.

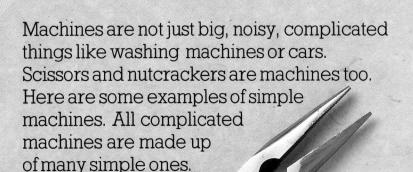

Levers

A lever is a stiff rod that turns (pivots) around a point called a fulcrum. It can produce large lifting or turning forces. A seesaw is a lever, and so are nutcrackers, pliers, scissors, clothespins and shovels.

Wedges

A wedge is the same shape as two slopes stuck back-to-back. When you push a wedge into a gap, you need a smaller amount of force to push two things apart. The blades of chisels, axes and knives are all wedges.

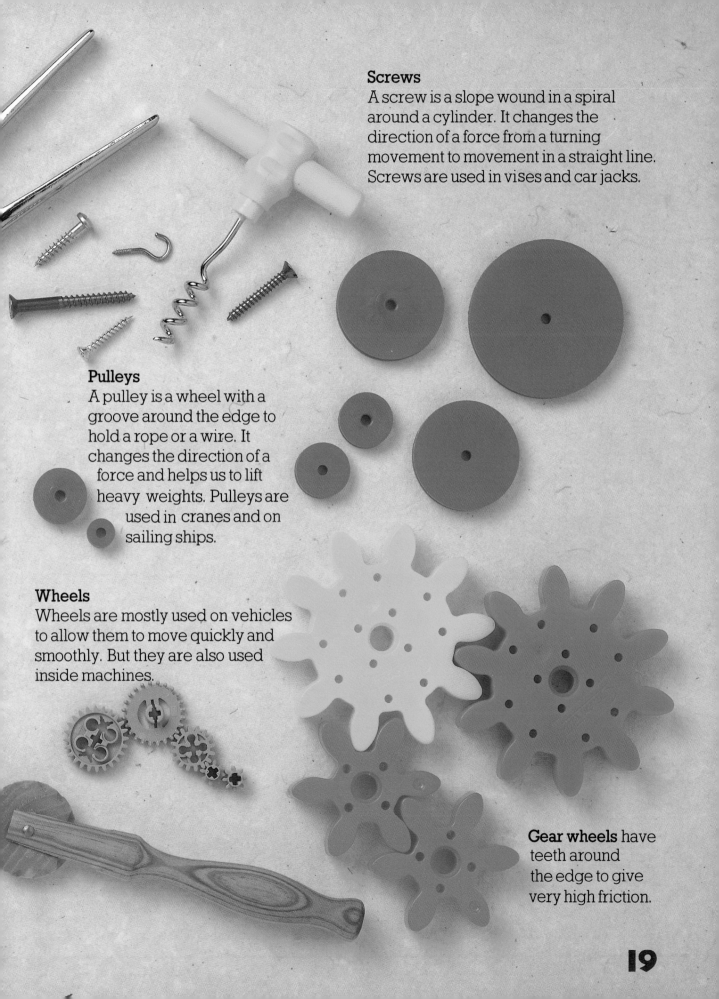

Screws

A screw is a slope wound in a spiral around a cylinder. It changes the direction of a force from a turning movement to movement in a straight line. Screws are used in vises and car jacks.

Pulleys

A pulley is a wheel with a groove around the edge to hold a rope or a wire. It changes the direction of a force and helps us to lift heavy weights. Pulleys are used in cranes and on sailing ships.

Wheels

Wheels are mostly used on vehicles to allow them to move quickly and smoothly. But they are also used inside machines.

Gear wheels have teeth around the edge to give very high friction.

LEVERS

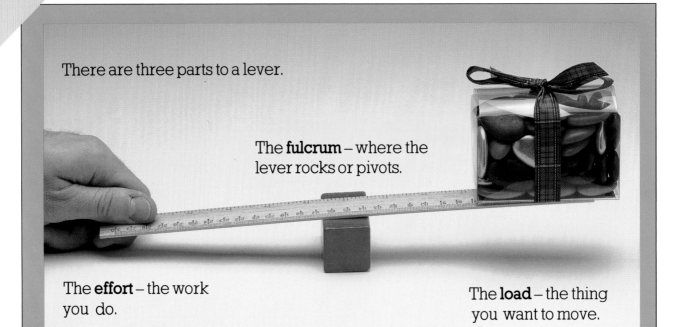

There are three parts to a lever.

The **fulcrum** – where the lever rocks or pivots.

The **effort** – the work you do.

The **load** – the thing you want to move.

To find out how levers work, try this test.

1 Balance a ruler on a small wooden block. Start with the block in the middle of the ruler.

2 Put a weight on one end of the ruler.

3 To lift the weight, push down on the other end of the ruler.

4 Now move the block nearer to the weight. When you try and lift the weight, you will need less pushing force.

5 Try moving the block nearer to your hand. What happens this time?

The shadoof is a type of lever used in Egypt to lift water from rivers and wells. Can you figure out where the load, fulcrum and effort are?

Each of these children is using a different kind of lever to do a different job. Can you find some more examples of each kind of lever?

Scissors

effort

fulcrum

load

Salad tongs

effort

fulcrum

load

Nutcrackers

fulcrum

load

effort

WHEELS

For thousands of years, people have used wheels to make their lives easier. But a wheel on its own is not much use. It needs something to turn on. This is called an axle. On toy cars and trucks, can you see how the wheels are fixed to the axles? The axle and the wheels can be fixed so that they both turn together. Or the wheels can turn while the axle stays still.

Try making wheels to match the shapes of these buttons. Which shape turns most easily? Round wheels roll over surfaces instead of sliding. This means there is less friction between the wheels and the surfaces.

Gear wheels are fixed in machines to make different parts move at different speeds or in different directions. They can also be used to multiply or reduce the power of machines. So, a gear wheel can turn a small force in one place into a larger force somewhere else.

The gear wheels in a clock are different sizes and are arranged to make the hands turn at different speeds. The big hand goes around once an hour and the small hand goes around once every twelve hours.

Try and fit different sized gear wheels together. When you turn a large gear wheel around once, how many times do the smaller wheels turn around? In which direction do the different wheels turn?

PULLEYS

Cranes use pulleys to lift enormous loads. The biggest cranes have three or four pulleys working together. With more pulleys, it is easier to lift heavy loads, but a lot more rope is needed. As with all machines, you do not get something for nothing.

Use two pulleys to lift a load.

1 Find two spools. Ask an adult to help you make a hook for each spool by bending some thick wire through the spools.

2 Screw two hooks into the top of a doorway. Hang a spool from one hook and tie a length of string or rope to the other hook.

3 Thread the string or rope around both spools as in the picture.

4 Hang a weight, such as a brick, on the bottom pulley and pull on the rope to lift the brick. You need only a small force to lift the heavy weight, but you have to pull on a lot of rope and it takes more time than just lifting the brick right off the floor.

SLOPES AND SCREWS

If you walk up a hill on a spiral path, it is much easier than going straight up. A screw uses the same idea. It is a long slope that has been "rolled up" so it takes up less space. To check this for yourself, cut out a paper slope and wrap it around a pencil.

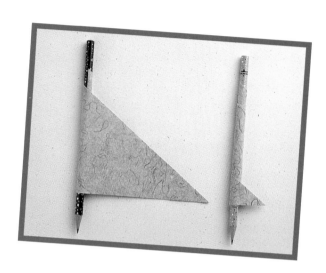

In the third century BC, the Greek scientist Archimedes invented a machine with a screw to lift water from the hold of a ship. This is called an Archimedes screw. Nowadays, some farmers in Egypt still use an Archimedes screw to lift water up to their fields.

These children are turning a model of an Archimedes screw.

BICYCLES

A bicycle is a well designed machine that allows us to move faster and more easily than we could by walking or running. If you look carefully at a bicycle, you will find examples of many of the forces and simple machines described in this book.

The handlebars steer and change the direction of movement. The rough surface on the handlebars helps you to grip.

Pulling on the brake levers makes the brakes push against the wheels.

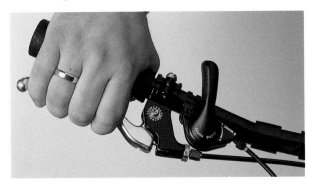

The muscles in your legs and feet provide the force to push the pedals around.

A pedal is attached to a metal bar called a crank. Ball bearings in the pedal crank reduce friction and help the wheels to spin smoothly.

The chain links the movement of the pedals to the wheels, which push the bicycle along.

The rough pattern on the tires is called the tread. It helps to increase the friction between the tires and the road and helps the tires to grip the road.

The brake pads rub against the wheels and friction makes the wheels stop turning.

The wheels make the bicycle roll along the ground instead of sliding. This reduces the friction and makes movement easier.

Gear-wheels change the force needed to start moving, stop moving or keep moving. Low gears help you put a lot of force on the back wheel to get started or to go up a hill.

MORE THINGS TO DO

Egg puzzle
How can you tell the difference between a raw egg and a cooked egg? Spin both eggs, then gently stop them from spinning and take your fingers away. The cooked egg will stay still, but the raw egg will start spinning again. This is because the liquid in the raw egg is still moving and makes the egg start spinning again.

Safety tests
Try these tests to find out more about the importance of wearing safety belts. To make a slope, prop up a piece of wood on some books. Put a large book or a brick at the bottom of the slope. Find a toy car and put an eraser on top of it. Roll the car down the slope. When the car hits the brick or book, the eraser should be flung over the top. How far does it go? To make the car go faster, make the slope steeper. How far does the eraser go now? Design a seat belt to hold the eraser firmly in place and repeat the test.

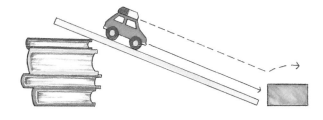

Gripping patterns
Use a wax crayon and some paper to make rubbings of all the different tread patterns you can find. You could try boots, bicycle tires, car tires and shoes.

Ball bearings
To find out more about how ball bearings reduce friction, find two cans that have a rim around the edge. Try spinning one can on top of the other. Now place some marbles around the rim of one can and spin the other can on top. The marbles make the can roll instead of sliding, which makes the movement much easier.

Looking for machines
Look around your home or school for examples of machines that use levers, wedges, screws, pulleys and wheels.

Design a vehicle
See if you can make a vehicle with wheels that will carry a load. The picture shows one design, but you can invent lots of others.

For the body, use a piece of cardboard, thin plywood, a box or a plastic container.

For the wheels, use wooden spools, plastic bottle caps, wood, cardboard or round plastic lids.

For the axles, use pencils, thin dowels or straws.

To hold the different parts together, use rubber bands, glue, modeling clay or thumb tacks.

Keep a record of the different stages involved in building your vehicle. You could draw sketches or take photographs.

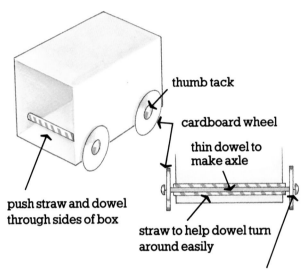

thumb tack

cardboard wheel

thin dowel to make axle

push straw and dowel through sides of box

straw to help dowel turn around easily

thumb tack in end of dowel to stop wheel falling off

Sliding down slopes

Tie a long length of string to the handle on a cupboard or chest of drawers. Tie the other end of the thread to a heavy object on the floor to make a steep slope. Make a hook from a paper clip and hang this on the string. Put different objects, such as buttons, more paper clips or a pencil sharpener on the hook and time how long each one takes to slide down the slope. Change the angle of the slope or make the slope out of different materials and repeat the test.

Make a pulley

Use four wooden spools and a long piece of rope or string to make a pulley that looks like the one in the picture below. Look back at page 24 for more information about making pulleys.

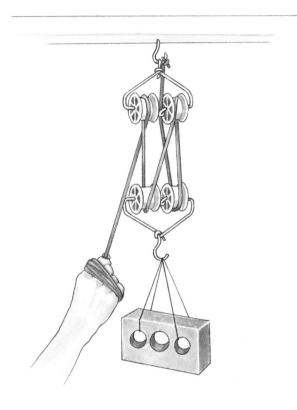

All about bicycles

Make a collection of different bicycles with your friends. How are the designs different? How heavy or light is each bicycle? Which bicycle is best for riding over rough ground? Which one goes fastest?

Test the brakes of a bicycle on different surfaces. How long does it take to stop on a wet playground? Or a dry playground?

Draw a picture of your bicycle and put an arrow in all the places where you think oil is needed.

DID YOU KNOW ?

▲ Prehistoric people probably used the heat from friction to start fires. They may have rubbed sticks or hit rocks together to make enough sparks to start a fire.

▲ The wheel was probably invented about 5000 years ago in Sumer, known today as Iraq. The first wheels were made of wood and the axles were fixed in position, which meant that the wheels could not turn left or right. A wheel can reduce the friction between a load and the ground by as much as a hundred times.

▲ Polar bears can walk fast over slippery ice because they have special "non-slip" soles on their feet. The soles have a thick pad of fat covered with thick, roughened skin and long, tough hair. Scientists have studied polar bears' feet to help them design boots for people such as farmers or fire fighters, who have to walk on wet or greasy surfaces.

▲ Geckos can run up and down vertical surfaces (even panes of glass) and can walk upside-down on ceilings. Underneath their toes are ridges of scales that contain millions of tiny "hairs." The "hairs" push into tiny bumps or dips in the surface. If you try to pull a gecko from a sheet of glass, the glass may break before the gecko lets go.

▲ It is possible to make a snowball because ice melts under pressure. When you press powdery snow together, some of the snow melts. This helps the flakes to stick together. When you stop pressing the snow together, the melted snow turns back into ice again.

▲ The Greek scientist Archimedes, who lived about 2000 years ago, invented the compound pulley. He is said to have amazed his neighbors by pulling a loaded, three-masted ship up onto the beach all by himself. He was able to do this using a system of compound pulleys.

▲ In sports, we can use our arms as levers. Some baseball players can throw the ball as fast as 90mph. At this speed, the ball goes from pitcher to batter in less than half a second.

A tennis racket allows you to turn your arm into an extra-long lever and hit the ball with considerable force. Your shoulder acts as the fulcrum, allowing your arm to pivot. The muscles in your upper arm deliver the effort to the ball, which is the load.

▲ When you push against a wall, you move backward, but the wall also moves away from you. This is hard to believe because the movement is too small to see or measure…but it's true! It is like the children on roller skates pushing against each other on page 5 of this book. By using roller skates, we can see the movement in both directions.

▲ The Earth's atmosphere presses down on everything on Earth. In fact, there is more than the weight of two pounds of air pressing down on every square inch at the surface of the earth. This includes the ground, the walls of a building and even your skin. Suction cups take advantage of this by making a partial vacuum under the cup so air pressure pushes the suction cup against the wall.

GLOSSARY

Axle
The rod on which a wheel turns.

Ball bearings
Small, steel balls used to help parts of machines move more easily.

Force
A push or a pull that makes an object move or change its speed or direction. It is measured in newtons.

Friction
A force that occurs when two surfaces rub against each other. Heat is produced when friction takes place. Friction tends to slow down objects or stop them moving.

Fulcrum (pivot)
The point where a lever pivots or turns. The fulcrum can be at different places along a lever depending on the kind of lever (see pages 20–21).

Gear wheel
A wheel with teeth or cogs around the edge that fits into another gear wheel or the holes in a chain.

Lever
A simple machine consisting of a bar that moves (pivots) around a fixed point. When a force is applied to one part of a lever, another part of the lever moves something.

Lubricant
A substance, such as oil, which is used to prevent two surfaces touching. A lubricant helps to reduce friction, wear, overheating and rusting.

Machine
Any artificial device that allows us to use less effort to do a piece of work.

Pressure
The amount of force acting on a certain area. It is measured in newtons per square meter or millimeters of mercury.

Pulley
A simple machine for lifting objects. It consists of a wheel with a groove around the edge to take a cable or rope. As the cable or rope moves, the wheel turns.

Screw
A simple machine consisting of a slope wrapped in a spiral around a central rod. It can change a turning force into a much greater straight line force.

Tread
The pattern of ridges on tires or the soles of shoes.

Wedge
A simple machine consisting of two slopes (inclined planes) stuck back-to-back to make a triangular-shaped block. It can be pushed into a gap to force two objects apart.

Newton's laws of motion

1 An object that is not moving will stay still unless an outside force acts on it. A moving object will keep moving at the same speed and in the same direction unless an outside force makes it change its speed or direction of movement.

2 An object speeds up or slows down in direct proportion to the size of the force acting on it and inversely proportional to the mass of the object. The change in speed occurs in the direction of the force.

3 For every action, there is an equal and opposite reaction.

INDEX

Additional photographs: Bryan and Cherry Alexander Photography 16 (t); The J. Allen Cash Photolibrary 4; courtesy of ARC Construction 13 (t); Alan Cork 25 (cr); David Exton/Trustees of the Science Museum LP061 14 (c), LP057 25 (cl); Chris Fairclough 8 (t); The Hutchison Library/Chris Parker 20 (b); courtesy of S Sacker (Claydon) Ltd 6 (t). 6 (c); Frank Spooner Pictures/P Nightingale 7 (b); courtesy of the Transport and Road Research Laboratory 9 (b); Janet Watson 15 (t); ZEFA 17 (b), 24 (t)
Picture researcher: Sarah Ridley